Forklift Operator Training

By Roger Jefferies

Copyright Roger Jefferies 2011

Table of Contents

Chapter 1 - Difference between a car and a forklift

Chapter 2 - Forklift Load Attachments

Chapter 3 - Daily Checklist

Chapter 4 - Surface Conditions

Chapter 5 - Load Composition

Chapter 6 - Pedestrian Safety

Chapter 7 - Narrow Aisles

Chapter 8 - Overhead Guard

Chapter 9 - Loading - Unloading

Chapter 10 - Speed Limit

Chapter 11 Refresher Training

Chapter 12 - Hands On Training

Chapter 13 - Review Process

Chapter 14 - Written test

Chapter 15 - Written test - Answer Sheet

Chapter 16 - Hands On Evaluation check List

Chapter17 - Basic Daily Checklist

Chapter 18 - Extensive Daily Checklist

The materials contained in this training manual are intended to be one element of your overall forklift training program. The packet must be enhanced with any unique details of your facility or equipment. It must also be presented by a knowledgeable trainer who is both familiar and experienced with the safe operation of a forklift. Proper guidance and hands-on training are essential in the overall training of your personnel. As such this training packet alone does not constitute a complete training program, but represents only one of several tools to be used in the proper training and evaluation of your operators.

You should retain a record of all training, evaluation and testing involving forklift operators. These records, along with any performance reviews of the forklift operator should be retained in the operators file. These files can be very useful should you be audited, or in reviewing the overall performance of your training program

Formal Training

(Powered Industrial Lift trucks)

Figure 1 - Photo created by Tynt22

Training Requirements

The Occupational Safety and Health Administration (OSHA) requires all forklift operates to be properly trained. This training must include a formal training segment, a hands on training segment as well as a hands on evaluation. This training must be administered by a person with knowledge and experience in the operation of forklifts.

This training will instruct trainees in the proper operations of forklifts. The training will also include site specific instruction dealing with the unique characteristics or requirements of the workplace in which the operator will be working. This may include ramps, inclines, uneven surfaces or chemical hazards which are present.

The trainee will be allowed to operate a forklift during the training process as long as the trainee has shown the ability to control the machine and basic knowledge of safety requirements. Trainee's will only be allowed to operate a forklift when such operation poses no risk to other workers, pedestrians or onlookers in the area. All forklift operators must be a minimum of eighteen (18) years of age. All trainees must be observed at all times by a qualified person(s).

All segments of the training process should be documented and remain on file for future reference and training requirements. Any operator who is involved in an accident or observed operating a forklift in an unsafe manner will be required to undergo refresher training. The performance of all operators will be reviewed and evaluated at a minimum of every three (3) years. If an operator receives a poor or unacceptable evaluation the operator will be required to undergo refresher training.

Chapter One

Difference Between an Automobile and a Forklift

A standard sit-down forklift with a lift capacity of 5,000 pounds can weigh over 9,000 pounds with no load. This is more than twice the weight of an average sedan. The weight of the forklift will have a direct effect on the operation of the vehicle in the areas of braking and center of gravity. Due to the weigh and sturdy construction of the forklift, collisions can cause extensive damage to buildings, product, or shelving.

A forklift steers with its rear wheels. This is far different from a standard automobile which steers with its front wheels. As a result, the forklift has a wide rear-end swing which must be accounted for. The rear-end swing can cause severe damage to buildings, products and shelving. Rear-end swing is also a leading cause of pedestrian injuries. Pedestrians may become pinned between the forklift and another object when the machine makes a turn. This is often caused by pedestrians who are unfamiliar with forklift operation or operators who are unaware of their surroundings. Forklift operators must always be aware of any pedestrians walking or working near the forklift. Pedestrians always have the right-of-way.

Forklifts are often operated in very close confines. It is not uncommon for a forklift to have only inches of clearance on both sides of the machine. These conditions require the undivided attention of the forklift operator.

The standard forklift will have a much higher center of gravity (COG) than the standard automobile. This condition is further exaggerated when the machine is carrying a load, or maneuvering with the load raised. Forklifts have a tighter turning radius than an automobile but the turns must be made at lower speeds to avoid tip-over accidents, which are a leading cause of operator fatalities.

<u>Truck Controls and Instrumentation</u>

The forklift steering is controlled by a standard steering wheel. To turn right the operator will steer to the right. To turn left, the operator will steer to the left. The pivot point for the steering is the front axle.

The standard forklift will have three (3) foot control pedals. The pedal on the far right will be the accelerator. The speed of the forklift is controlled by depressing or releasing the accelerator pedal. The other two pedals are both brake pedals. Under normal operation the center brake pedal is the primary brake pedal. At times it will be necessary to apply pressure to the accelerator while also depressing the brake pedal. When this is the case, the left brake pedal will be used in conjunction with the accelerator pedal.

At times it will be necessary to increase the engine above an idle in order to raise a load. This can be done while the forklift is in gear. Depress the brake pedal and press down on the accelerator. Slowly release the brake pedal to allow the machine to inch forward as the load is simultaneously raised to the required height.

The emergency brake for the forklift will be located either on the dash of the forklift, or will be a smaller pedal on the floorboard of the machine. The emergency, or parking brake, must be disengaged when the machine is in operation. When the machine is parked or otherwise removed from service the emergency/parking brake must be engaged.

There will be a lever located on the left side of the steering column. This is the directional control. In order to move forward the lever must be pushed up/forward. To move the forklift in reverse pull the lever fully backward/down. The center position is neutral. When the machine is parked or taken out of service the directional control must be in the neutral position.

The three levers on the right side of the steering column control the mast functions.

A single lever will control the raising and lowering of the mast. To raise the mast/load the lever must be pulled back. To lower the mast/load the lever will be pushed forward. The speed at which the mast/load moves will be determined by the degree to which the lever is moved.

A second lever controls the tilt of the mast/load. In order to tilt the mast/load forward the lever should be moved forward. In order to tilt the mast/load back the lever needs to be moved back.

The third lever will allow the forks to be shifted to the left or right for more precise load placement. When traveling the forks/load should be centered . (Note: All forklifts do not have this side-shift option.)

Engine / Motor Operation

Forklifts are powered by a number of different engine/motor types. This can include a gasoline engine, diesel engine, LP gas engine, or Electric. Each engine/motor type will require different refueling techniques and each will have different maintenance issues and requirements.

Visibility

Forklifts have limited forward visibility due to the mast and hydraulic system. The location and size of the mast and its various components creates a number of blind spots when traveling forward. Vision can be further impaired when transporting a large or high load. If forward visibility is obstructed by a load, the forklift should be driven backwards. When operating in this manner the forklift should be operated at a reduced speed. The operator must always face in the direction of travel.

Authorized to Operate Forklifts

Only properly trained operators will be allowed to operate a forklift. Operators must be a minimum of eighteen (18) years of age and have completed all training segments. Operators must be certified on each type of machine they are to operate.

Persons involved in a training program are allowed to operate a forklift. Such operation will be limited to training purposes and will take place under conditions which do not present a danger to the operator, trainer, or other individuals in the area.

Chapter 2

Forklift Load Attachments

Forklifts can be adapted to a variety of working conditions and load handling characteristics. These may include optional load handling equipment replacing the standard forks. These attachments often require the operator to use different load handling procedures. Operators must be properly trained in the operation of each load handling attachment.

Slip Sheet - This attachment replaces the forks with a flat metal plate which slides under a cardboard sheet which is used in place of a wooden or plastic pallet.

Clamp Attachment - In place of forks the forklift will be equipped with two large clamps. These clamps close on each side of the load to be transported. The load is held securely in place by the clamps without the need for pallets or slip sheets. These attachments are often used for bales or rolls of material such as paper.

Barrel Clamp - This attachment is used specifically for the safe movement of barrels or drums, generally fifty-five (55) gallon drums, but other drum sizes can also be transported. This attachment is similar to the clamp attachment as it closes around the outside of the barrel to secure it in place.

Other attachments are available for specific operations. With each attachment the operator must receive proper training and orientation prior to using the equipment. All attachments must be installed by a competent professional in accordance with manufacturer specifications.

When adding or changing a load handling attachment, the weight of the forklift is also changed. This will have an effect on the load capacity and handling characteristics of the machine. All of these must be taken into consideration when operating the equipment. Never install an attachment on a forklift for which it is not intended.

Forklift Capacity

A forklift operates on the fulcrum principal. The fulcrum on the standard forklift is the front axle of the machine. The load capacity of the machine will be determined by the weight of the machine behind the fulcrum point. The capacity of each forklift will be listed on the data plate. These capacities should never be exceeded.

The listed load capacity of a forklift is based on the center of the load being twenty-four (24) inches from the backrest of the machine. As the center of the load moves further away from the backrest the lifting capacity of the forklift decreases. This information is also listed on the forklift data plate.

If a forklift has a lifting capacity of 5,000 pounds with a twenty-four (24) inch load center, this capacity will be reduced as the load center changes, or as the load is elevated. In some cases the load capacity can drop by fifty (50) percent or more when the load is elevated. This must be taken into account when handling any load. On the chart below you can see a machine capcity of 5,000 pounds is reduced to 2,000 pounds when the load is raised to a height of 169 inches or just over 14 feet.

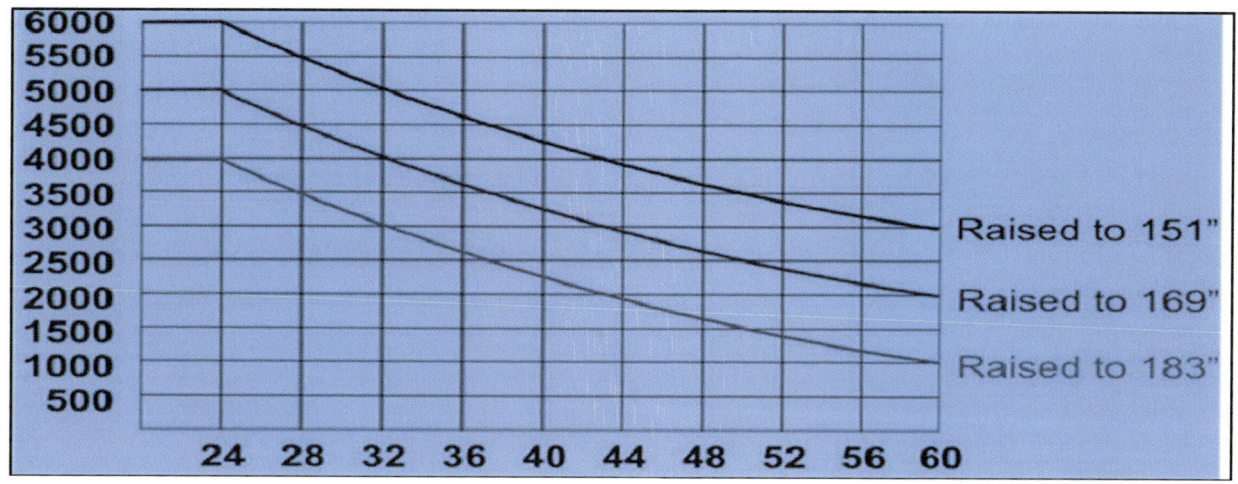

You may never add additional weights to a forklift in order to increase its lifting capacity.

Forklift Stability

The stability of a forklift is based on a stability triangle. The base of the triangle is the front axle. To form the remainder of the triangle draw a line from each front wheel to the center of the rear axle. This is the stability triangle and it should be kept in mind at all times. When the machine has no load, the center of gravity (COG) is located in the center of the triangle approximately at the location of the machine operator.

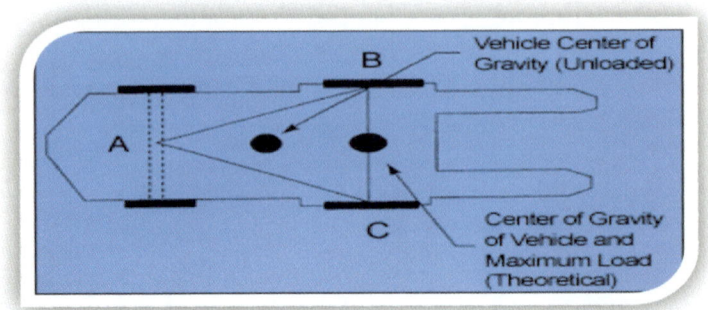

The COG shifts to the base of the triangle, or the front axle, when the machine is lifting a load. As long as the forklift is not turning, the COG will be located in the center of the front axle when supporting a maximum load.

As the machine turns, centrifugal force acts on the machine and will shift the COG away from center. The sharper the turn or more radical the movement of the machine, the further the COG will be shifted. If the COG ever shifts to a point it passes beyond the stability triangle the machine will become unstable and in danger of tipping over.

Centrifugal force can act on a forklift without a load as well as a fully loaded machine. If traveling forward a hard right turn will cause centrifugal force to pull the machine to the left. All turns must be made at slow speeds to maintain a proper COG.

Stability Triangle

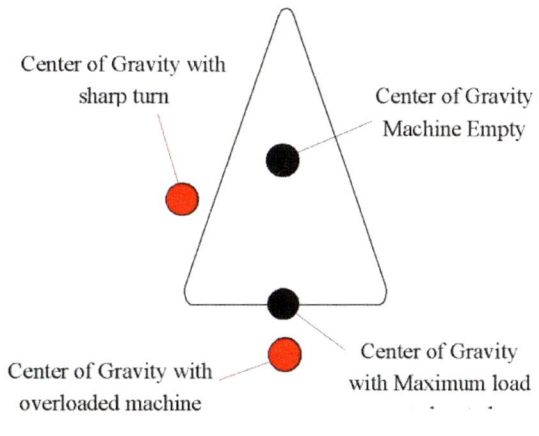

The COG will also be affected by inclines or uneven working surfaces. When operating on an uneven surface the forklift must be operated at reduced speeds and loads must be kept as low as possible, generally no more than three to four (3 - 4) inches from the ground. When on an uneven surface, the COG will shift very rapidly if the load is elevated or the machine is turned abruptly.

Chapter 3

Daily Checklist

OSHA requires a forklift check list be completed at the beginning of each shift for all machines in operation. Ideally this checklist will be completed by the operator who will be on the machine being inspected. The check list, often referred to as a daily check list, contains a number of items which must be inspected. This is to insure the machine is safe and ready to be used. The checklist must be completed prior to the machine being put into operation. The checklist must then be filed and maintained for a period of five (5) years.

All checklist are not the same. Some employers desire to have all elements of the forklift checked prior to it being operated. In these instances the checklist can be very extensive. Other employers wish to only have the key components checked to insure the machine is safe to operate.

If any problems are discovered they must be noted on the checklist. In addition, these problems should be brought to the attention of a supervisor or a member of the maintenance department prior to the machine being put into operation. At no time should an operator put a machine into service which they feel is unsafe or is in need of service.

Refueling the Forklift

Each machine type (LP, gasoline, electric) will have different refueling requirements and safety procedures. In all cases there is no smoking in the refueling area. The forklift engine must be turned off. At all times the proper Personal Protective Equipment (PPE) must be worn.

LP Gas machines; Eye and hand protection must be worn when refueling LP machines or exchanging LP tanks. LP gas is extremely cold and can cause severe burns. If exchanging an empty tank for a full tank be certain the valves on both tanks are in the closed position before handling them. Do not refuel the machine or exchange tanks near any open flame or other potential ignition source.

Gasoline machines; Eye protection is advised and refueling should take place well away from any open flames or ignition sources. Do not overfill the container and be certain you are grounded before beginning the refueling process. Gasoline fumes can be harmful, stand away from the unit while the refueling is taking place, but do not leave the forklift unattended while refueling.

Electric machines; Make certain both the forklift and the charger are turned off. Make certain the battery on the forklift and the charger to which it is be connected are the same voltage. In most cases the battery should not be placed on the charger until the battery is discharged to fifty (50) percent or lower. Make certain the connection is secure, then turn the charger to daily or normal charge. Approximately once a week the charger should be set to weekend or equalize charge. Management or the maintenance department will determine this schedule.

Operating Limitations

Forklifts must be operated at safe speeds at all times and the load capacity of the forklift may never be exceeded. If operated at high speeds or with excessive loads the forklift can become unstable and tip over. A tip over accident is the leading cause of injury to forklift operators and is often fatal. If the machine begins to tip over there are specific procedures you must follow to reduce the chance of serious injury or death.

In case of Tip Over;

DO NOT JUMP - you are far safer remaining in the operator's compartment. Always wear your seatbelt to help you remain in the seat at all times. Operators have been killed when they attempted to jump from the forklift only to be crushed by the overhead guard.

Hold on tightly to the steering wheel. This will not only brace you against the impact, but will help insure your hands and arms remain inside the operator's compartment.

Brace your feet. Spread your feet as wide as possible but be certain they remain inside the operator's compartment. This will help steady you and protect you during the impact.

Lean away from the impact. If the machine is tipping to the left, lean to your right. This will help prevent you from striking your head on the ground during impact.

Lean forward. This will also help protect your head during impact.

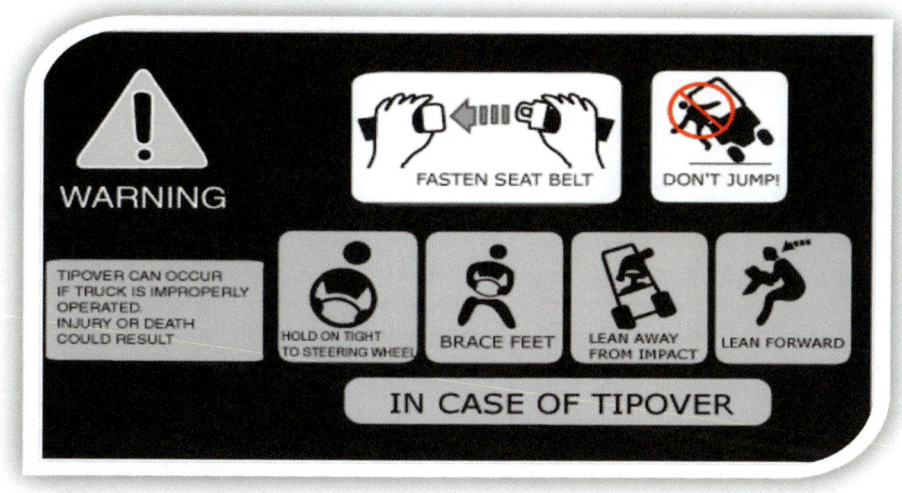

Chapter 4

Surface Conditions

Forklift operators must always take into account the different types of surface conditions. These may include smooth concrete, wood, gravel or asphalt. Extreme care should be taken when operating a forklift outdoors where the terrain can be rough. Potholes or uneven terrain can cause the COG to shift and the forklift to become unstable.

Any raised or elevated platforms must be of sufficient strength to support the weight of the forklift, operator and load combined. If in doubt, the machine should not be operated on any elevated or raised platforms.

Wet or damp surfaces may be encountered in many facilities. Most common are wet loading dock areas where rain or snow has blown into the building through open dock doors. Conditions such as extreme humidity or fog can cause floors to become damp and extremely treacherous. Most forklift tires are not designed to grip wet surfaces and the forklifts will become difficult to control and stop. Operators must reduce speed and corner carefully when operating on wet or damp surfaces.

Receiving/Shipping Docks

Shipping docks pose unique risks to forklift operation. These areas are often crowded and very busy. When operating in these areas operators should reduce speed, be very attentive of pedestrian traffic and sound their horn more frequently.

Open dock doors pose a serious risk to forklift safety. Extreme care must be used when operating near dock doors which are open and unoccupied by a truck or other vehicle. Should a forklift fall from the dock door the operator is in danger of serious or fatal injury. If possible, all unused dock doors should be closed or otherwise blocked to any type of vehicle traffic. No forklift should be operated within six (6) feet of any open dock door.

Chapter 5

Load Composition

The size and shape of the load being handled can be as important as the weight of the load. As discussed in a previous section, the load center of the standard forklift is set at twenty-four (24) inches from the backrest. If this load center shifts further from the back rest the load capacity of the forklift is reduced. The load capacity will also be reduced further as the load is elevated.

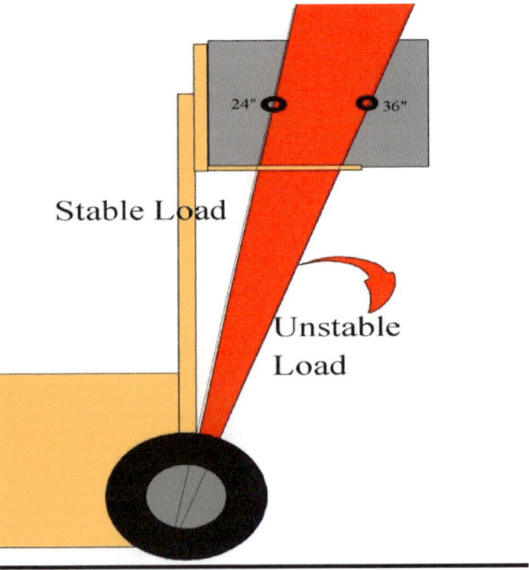

The load center will also change as the mast is titled forward or back. If the load is titled forward the load center moves further from the backrest and the load capacity is reduced.

Whenever possible loads should be level and spread evenly from side to side. High loads should be allowed to rest against the backrest for added stability as well as keeping the load center and COG as near center as possible.

Load Stability

Loads must be secured. The movement of the forklift as well as the roughness of the working surface will cause vibration. The higher the rate of speed, the more severe the vibration. This vibration can cause the load to shift and become unstable. Loads must be secured before being transported. Falling loads can result in damage to product, machinery and shelving. Falling loads can also result in the injury of pedestrians or other workers in the area.

Load Manipulation

The size and configuration of the load being transported will partially dictate the maximum speed of the forklift. Loads will shift as the forklift makes turns when centrifugal force takes effect. If turns are made too rapidly, or too abruptly, the load can shift and spill. This poses a risk to pedestrians in the area and can result in damage to product and facilities as well. Speeds must be reduced when transporting loads, especially when cornering.

When transporting a load the load must be kept as low as possible. This should be three to four (3 - 4) inches from the ground. The load should also be allowed to rest against the back rest and tilted back slightly. This will help stabilize the load when braking.

Accelerate and brake evenly. Rapid starts and stops cause loads to shift and fall. Reduce speed when cornering to avoid spilling loads.

When placing a load in shelving or pallet racks, approach the rack slowly. Keep the load as low as possible until you are properly aligned with the slot in which you intend to insert the load. Raise the load slowly, keeping it titled back slightly. Allow the load to clear the bottom of the shelf on which it is too be placed before moving forward. Before moving forward be certain to check the overhead clearance to be certain the load will not contact the next shelf. Move forward slowly until the load is in place. Tilt the load forward and gently lower it until it rests on the racking. Level the forks and slowly back out. Do not maneuver until the forks have exited the pallet.

Before placing any load in a shelf or pallet rack the operator must be certain the load does not exceed the capacity of the shelf/racking. No loads should be placed in damaged racking. Watch for rear end swing when moving into position or backing away from racking.

Load Stacking/Un-stacking

In many instances loads which are placed on pallets can be double stacked, one pallet load placed atop another. This can only be done when the base pallet will support the weight of the additional pallet atop it. The top of the load must also present enough of a platform to provide stability for the second pallet. In most cases pallets should not be stacked higher than two levels. When stacking pallets all pallets must be free from damage.

No loads will be placed in such as position or location that they will block or impede access to exits, fire equipment, or medical equipment such as first aide cabinets or AED's. Loads should be located in areas designed for these purposes.

Chapter 6

Pedestrian Safety

No one should ever be allowed to walk beneath the raised forks of a forklift, loaded or unloaded.

Forklifts will not be driven up to a person who is standing at a workstation, wall, or other solid object.

Forklift operators should not talk with pedestrians while operating the forklift. If a conversation must take place, the load should be lowered to the floor, the engine taken out of gear, and the brake on.

Forklifts must slow when approaching a pedestrian crosswalk. The operator should sound the horn when approaching the crosswalk.

When operating near any pedestrians the forklift will be operated at reduced speed. Pedestrians should not be allowed near the rear of the forklift. A leading cause of pedestrian injury and death results from the pedestrian being struck or pinned when the rear of the forklift swings out during a turn.

Pedestrians have the right-of-way.

Forklifts must slow or come to a complete stop when encountering a blind corner or intersection.

Riders

At no time are any riders allowed on the forklift.

Persons are never to be raised on the forklift forks or other attachments.

Persons may be raised in an approved basket which is properly enclosed with guard rails and lockable gate.

The cage being used to raise the person will be secured to the forklift mast with either a safety chain or cable.

The cage will be equipped with a guard a minimum of seventy-two (72) inches in height which will prevent the person from inserting their hand or arm into the mast of the forklift.

The operator must remain at the forklift controls while another worker is raised on the work platform.

Chapter 7

Narrow Aisles

Forklifts are often required to operate in close confines and narrow aisles. When this is the case the forklift will be operated at reduced speed. No pedestrians or workers may be in the aisle when the forklift is working in the aisle.

Restricted Work Environments

Certain work environments contain dangers which do not allow standard forklifts to be used. This can include areas such as grain elevators, paint booths, or areas where other chemicals or substances are present.

[Trainer: Cover the specific areas of concern within your facility.]

Overhead Restrictions

Each workplace will present different conditions which the forklift operator must constantly take into consideration. These conditions can also change within a single facility, even moment to moment. Overhead obstructions such as lights, conduit, sprinkler lines, duct work, guide wires, electrical wires/cables and gas lines must be accounted for at all times.

If the forklift must be operated in confines which are too close for normal operations, a spotter should be used. This person will help guide and direct the forklift around and away from obstacles.

The person spotting for the operator must remain a safe distance away from the forklift and can never move or stand beneath a raised load, or raised forks.

Chapter 8

Overhead Guard

The forklift will be equipped with an overhead guard which is intended to protect the operator from falling objects. This guard is intended to protect against small objects and may not be rated to support a load equally the lifting capacity of the forklift.

Any damage to the overhead guard should be reported to management immediately.

Backrest

A backrest, or carriage, will be mounted to the fork assembly. The backrest is used to stabilize the load and help prevent spillage. The backrest also helps prevent a shifting load from damaging the hydraulics and mechanics of the mast.

The backrest must remain in place at all times. It not only protects the mechanical portions of the forklift but also provides protection for the operator from falling items when the load is being elevated and placed into racking or shelving. Any damage to the backrest must be reported to management immediately.

Railroad Tracks

Forklifts must always cross railroad tracks diagonally and at greatly reduced speeds. The rough terrain of the railroad tracks can cause the COG to shift and can also affect the steering of the unit.

In many locations forklifts are required to come to a complete stop prior to crossing railroad tracks.

No forklift will be parked within 8 feet of the center point of any railroad track.

Driving Restrictions

Horseplay cannot be tolerated. Any operator who engages in horseplay while operating a forklift will be required to undergo refresher training. In many facilities the forklift operator is forbidden to operate a forklift until they have successfully completed refresher training.

(When following another forklift a minimum of three truck lengths will be maintained between the two forklifts.

Elevators

Elevators are to be approached slowly. Before entering the elevator the operator must insure adequate overhead clearance for both the forklift and any load being transported.

The operator is responsible for ensuring the elevator's lift capacity is sufficient to lift the forklift and any cargo being transported.

Once on the elevator the forks will be completely lowered, the engine turned off and taken out of gear, and the parking brake set to prevent movement.

Chapter 9

Loading/Unloading Trucks/Containers

When loading or unloading a truck or container the forklift operator must be certain all preparations have been made and all precautions are taken. These include;

The wheels of the truck are properly chocked.

If there is no tractor attached to the trailer a nose jack should be placed under the front of the trailer to prevent tipping.

A procedure is in place to be certain the trailer is not moved while the loading/unloading process is taking place.

The trailer is secured to the dock via a locking system.

Light system is used to indicate when the process is completed.

The driver is informed when the process begins and when it is completed. In some cases the driver must remain in the building until the process is completed.

The forklift operator must inspect the floor of the trailer/container to be certain it will support the weight of the forklift and the intended load. If the floor is damaged or otherwise appears inadequate they must notify management before proceeding.

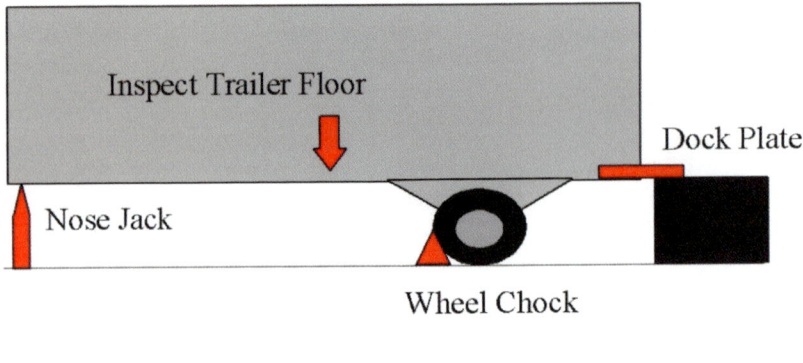

The forklift must be operated at reduced speed while moving in or out of a trailer and while inside the trailer/container.

Loading Ramps

When loading or unloading trucks a loading ramp is used to span the distance between the dock and the rear of the truck. This ramp must be of sufficient strength to support the weight of the forklift, the operator and the load being transported.

The loading ramp must be anchored or otherwise secured to prevent it from slipping when the forklift passes over it. This is normally achieved by a pole or center section of the ramp extending downward between the truck and dock.

Inclines

A forklift should never be parked or left unattended while on an incline.

When driving on an incline the forklift should be driven straight with no turns or curves. This will help insure the COG remains within the stability triangle and reduce the risk of tip-over.

The forklift must be backed down the incline with the load on the uphill side.

The forklift must be driven up the incline with the load on the uphill side. {Remember the rule: Drive-Up and Back-Down}

Chapter 10

Speed Limit

OSHA does not specify a speed limit for forklift operation. The OSHA guidelines do require that a forklift be operated at a safe speed at all times. Speed must be reduced when cornering, operating on inclines, or on wet or damp surfaces. Speeds must also be reduced when transporting loads or when near pedestrians.

Parking

Forklifts should be parked in an approved area only and should never be placed in a position in which it will block or inhibit access to exits, fire-fighting equipment, or medical equipment.

When a forklift is parked it must;

Be taken out of gear.

Key in the 'off' position.

Parking brake set.

Forks lowered to the floor.

Keys removed to prevent unauthorized use.

A forklift operator may exit the forklift and leave it running only if;

Taken out of gear.

Forks are lowered to the floor.

Parking brake is set.

Operator must remain within eye-sight of the forklift.

Operator must remain within 25 feet of the forklift.

Chapter 11

Refresher Training

All forklift operators are required to undergo an evaluation of their performance and driving skills a minimum of every thirty-six (36) months. If any concerns are noted the operator must undergo refresher training in order to maintain their certification.

A forklift operator will be required to undergo refresher training if;

They are involved in an accident or near miss incident.

They are observed operating the forklift in an unsafe manner.

They receive an evaluation indicating their operator skills are less than adequate.

They are assigned to a new type/style of forklift.

There are changes in the work environment which may affect the safe and routine operation of forklifts.

Chapter 12

Forklift Training

Hands-On Training

(Powered Industrial Lift trucks)

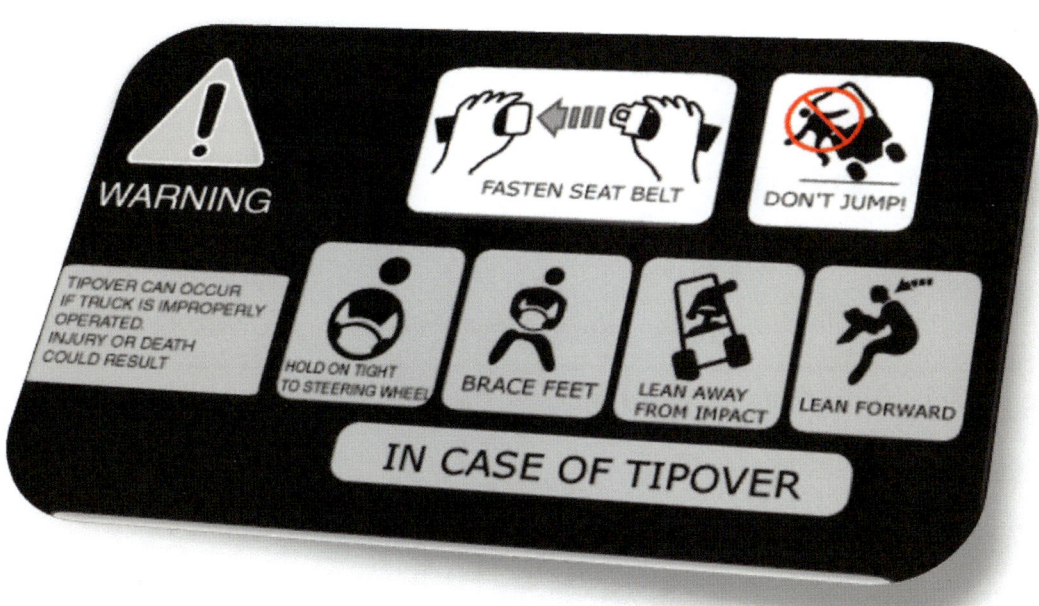

Hands-On Training

Daily Check List

Trainee will be shown where the checklists are kept, the proper method in which to complete the checklist and where to deposit the checklist when completed.

Items to be checked before starting the forklift.

Inspect tires for excessive wear or separation.

Inspect hydraulic hoses for cracking or leaks.

Walk around the forklift and look for any obvious damage.

Make certain the forks are locked into position and are free of damage.

Check all controls - move the controls and release them, they should return to the neutral position.

Sound the horn.

Items to be checked with the motor running.

Raise and lower the forks, check for smooth operation.

Check the tilt and side-shift functions for smooth operation.

Check the parking brake - With the brake engaged, put the forklift into gear. The machine should not move.

Check any lights on the forklift.

Drive forward slowly and test the brakes. The machine should stop smoothly.

Check for excessive play in the steering. The steering wheel should not move more than one (1) inch before the machine responds.

Check where the machine was parked for any signs of leaks.

Some check lists will require additional items to be checked such as oil levels, coolant levels, and hydraulic levels. All items on the checklist must be completed before the machine is to be put into service.

Review the Forklift

Trainee must be shown the crucial elements of the forklift which they will be dealing with on a daily basis. The more experience the operator has the more they will have a feel for the forklift and be able to tell when a problem exists.

The following areas must be covered in enough detail to allow the trainee a firm understanding of the machine.

Adjusting the forks - locking the forks in place.

How to open the engine compartment and examine primary engine components - oil, water, transmission fluid, battery, hydraulic fluid.

Proper method to secure the engine compartment and operator's seat.

How to properly adjust the seat, steering column and other controls to properly fit the individual operator.

Point out the location of the data plate and how to interpret the information presented.

Review any special features of the forklift, their location, and their proper use. This may include a fire extinguisher, PPE, and mirrors.

Starting the Forklift

The trainee will mount the forklift and prepare to operate the forklift. This will include adjusting the seats, mirror, steering column, and seat belt. After starting the forklift the trainee will be shown the basics of the machine operation.

At a minimum the following components will be covered in enough detail to allow the trainee to be comfortable and have a reasonable confidence in the proper operation of the forklift controls.

Hydraulic controls - raise and lower the mast, tilt the mast forward and back, and shift the fork carriage side to side.

Trainee will be shown how to visually check the proper traveling height of the forks, which will be 3 to 4 inches from the ground.

How to engage and dis-engage the emergency parking brake.

Re-Fueling the Forklift

Trainee will be shown the proper method and location to re-fuel or recharge the forklift. The following items must be covered in enough detail to allow the trainee to become comfortable and familiar with the proper refueling method for the forklift on which the trainee will be operating.

What personal protective equipment (PPE) must be worn and where this equipment is located.

Where the refueling process is to take place.

How to properly access the refueling equipment.

Proper preparation and filing of any documentation required when refueling a forklift.

Location of emergency equipment such as fire extinguisher, emergency eye wash station, first aid cabinet.

Proper method to make all connections involved in re-fueling the forklift.

Proper method of reporting any emergency situations, spills, or mechanical issues encountered while the refueling is in process.

Basic Travel

A trainee may only operate a forklift when such operation will not endanger pedestrians, co-workers, the trainer, or the trainee. First time operators should operate forklifts in a remote area away from personnel and which allows enough open space to avoid contact with racking, walls, or other materials.

Operator will drive the forklift at a reduced rate of speed until they have demonstrated a familiarity with the machine and its controls.

Trainee will be allowed to operate the forklift in a series of turns, starts and stops to become familiar with the stopping distance and acceleration of the machine. The trainer will discuss any errors at the time they are observed. Trainer will demonstrate how to determine turn radius and turn times for the forklift. The trainee will then demonstrate the ability to properly control the forklift in turns traveling both forward and in reverse.

Trainee will not move to the next portion of the training until the trainer is satisfied the trainee has a firm understanding and ability of how to control the forklift in a safe manner.

Load Handling

Trainee will be provided with a sample load (pallet) with which to work/practice. The practice pallet should be well below the maximum weight capacity of the forklift, but of sufficient mass to allow a feel for how the forklift handles and operates with a load.

The trainer will instruct the trainee on the proper method to approach the load and raise it to travel height of three to four (3-4) inches off the floor. The trainee will then be observed traveling and making turns while transporting the load. Trainee will set the load down and pick it up several times until the trainer is certain they have a good feel for the controls and method of approaching the load.

Load Placement

Once the trainee has shown their ability to safely drive and operate the forklift in an open area they will move on to shelving or pallet racking areas.

The trainer will demonstrate the proper method of placing a loaded pallet in the pallet racking. The trainee will observe and ask any questions prior to attempting the procedure themselves. Once comfortable with the knowledge, the trainee will take the controls and place the pallet in the pallet racking as instructed by the trainer. They will then drive away from the pallet racking before returning to retrieve the pallet from the pallet rack.

The trainer will observe the trainee as they place the load in the racking and remove it. As the trainee progresses the trainer will have the trainee place the pallet in a more difficult or higher racking locations. This practice will continue until the trainer and trainee are both satisfied the trainee has sufficient skills to safely handle load placement and removal.

Parking the Forklift

Review proper procedures for parking the forklift. Show examples of areas where the forklift should not be parked (exits, fire equipment) and the preferred area for forklift parking

Procedures for parking a forklift.

Forks lowered completely to the ground.

Engine taken out of gear.

Engine turned off.

Parking brake set.

Remove keys. (Not a policy with all facilities.)

Trucks / Loading Docks

The trainee will be shown the proper method to prepare a truck for loading or unloading. These procedures include;

Chock the wheels of the trailer to prevent movement.

If no truck is present, place a nose jack under the front of the trailer to prevent tipping.

Proper placement of the dock plate, including securing it in place.

Procedures for making certain the trailer is not pulled away until the loading/unloading process is complete.

Inspection of the trailer floor for weakness or damage.

The trainer will observe the trainee's movements in and out of the trailer and how the load is being handled and placed. The trainer will offer immediate input and correct any errors observed.

The trainee will be shown the proper method to place the pallets in the trailer and to secure the load for transport when required.

All pages beyond this point may be cut out and reproduced as required to complete your training.

Chapter 13

Forklift Training

Formal Training - Written test

Multiple Choice Questions - Four points each

1. How often must a check list be completed on a forklift?

____ a) Daily

____ b) Weekly

____ c) Beginning of each shift

____ d) When a problem is suspected

2. A forklift is different from a standard automobile in what way?

____ a) Steers with rear tires

____ b) Weight

____ c) Visibility

____ d) None of the above

____ e) All of the above

3. At what height should a load be carried under normal conditions?

____ a) Under a foot

____ b) 6 - 8 inches

____ c) 3 -4 inches

____ d) Operator's discretion

4. If a load is high enough to block the operator's view, they should?

____ a) Reduce speed and move forward

____ b) Sound horn and move forward slowly

____ c) Drive the machine in reverse

5. What is the leading cause of operator injury and death?

____ a) Rear-end swing

____ b) Falling Objects

____ c) Tip-Over Accidents

____ d) Chemical Exposure

____ e) None of the above

6. What can cause the lift capacity of a forklift to change?

____ a) Load being elevated

____ b) Changing the load center

____ c) Tilting the load

____ d) Changing lifting attachment

____ e) All of the above

7. What is a leading cause of pedestrian injuries?

_____ a) Tip-Over accidents

_____ b) Falling objects

_____ c) Rear-End Swing

_____ d) Falling from loading dock

_____ e) None of the above

8. When operating near an open dock door, what is the minimum safe distance to remain away from the dock door?

_____ a) 2 to 3 feet

_____ b) 10 feet

_____ c) 3 to 4 feet

_____ d) 6 feet

9. When crossing railroad tracks, a forklift should be driven across the tracks in what direction?

_____ a) In reverse

_____ b) Diagonally

_____ c) Forward, but slowly

_____ d) Operator's choice

10. Persons may be raised in an approved work basket if?

_____ a) The operator remains at the controls at all times

____ b) The basket is secured to the forklift

____ c) The person is a member of the maintenance department

____ d) If the forklift is inside the building only

____ e) Both a and b

____ f) Both c and d

True or False Questions - Four points each

1. ____ True ____ False - Riders are never allowed on a forklift.

2. ____ True ____ False - All forklifts has the same lifting capacity.

3. ____ True ____ False - The OSHA speed limit for forklifts is 5 m.p.h.

4. ____ True ____ False - Inclines do not affect the center of gravity of a forklift.

5. ____ True ____ False - Pedestrians must yield to forklifts.

6. ____ True ____ False - Operators must always wear their seatbelts.

7. ____ True ____ False - A forklift cannot be parked closer than 8 feet from the center of a railroad track.

8. ____ True ____ False - The load capacity of a forklift is based on a 24 inch load center.

9. ____ True ____ False - Centrifugal force has no effect of the forklift's center of gravity.

10. ____ True ____ False - Forklifts weight enough that wet surface conditions do not effect them.

Fill In The Blank Questions - Four points each(one point per blank)

1. If a forklift is tipping over, the operator should not jump. In addition, to further protect themselves the operator should;

   a)_____

   b)_____

   c)_____

   d)_____

2. When parking a forklift the operator must be certain what steps are taken?

   a)_____

   b)_____

   c)_____

   d)_____

3. When operating on an incline the forklift operators should reduce _____ and make no _____ to prevent the center of gravity from shifting. The forklift should be _____ up the incline, and _____ down the incline.

4. When loading or unloading a truck/trailer the forklift operator must complete the following preliminary procedures. They must insure the wheels of the trailer are _____ and place a _____ _____ under the front of the trailer if no truck is attached. Inspect the trailer _____ to be certain it will support

the weight of the forklift and the load being carried. Be certain the dock plate is

_____ before driving over it.

5. Forklifts can be powered by a number of different types of engines/motors. These include;

a)_____

b) _____

c) _____

d) _____

Chapter 14

Forklift Training

Formal Training - Written test - Answer Sheet

Multiple Choice Questions - Four points each

1. How often must a check list be completed on a forklift?

   ____ a) Daily

   ____ b) Weekly

   _X_ c) Beginning of each shift

   ____ d) When a problem is suspected

2. A forklift is different from a standard automobile in what way?

   ____ a) Steers with rear tires

   ____ b) Weight

   ____ c) Visibility

   ____ d) None of the above

   _X_ e) All of the above

3. At what height should a load be carried under normal conditions?

____ a) Under a foot

____ b) 6 - 8 inches

_X_ c) 3 -4 inches

____ d) Operator's discretion

4. If a load is high enough to block the operator's view, they should?

____ a) Reduce speed and move forward

____ b) Sound horn and move forward slowly

_X_ c) Drive the machine in reverse

5. What is the leading cause of operator injury and death?

____ a) Rear-end swing

____ b) Falling Objects

_X_ c) Tip-Over Accidents

____ d) Chemical Exposure

____ e) None of the above

6. What can cause the lift capacity of a forklift to change?

____ a) Load being elevated

____ b) Changing the load center

____ c) Tilting the load

____ d) Changing lifting attachment

_X_ e) All of the above

7. What is a leading cause of pedestrian injuries?

   ____ a) Tip-Over accidents

   ____ b) Falling objects

   X c) Rear-End Swing

   ____ d) Falling from loading dock

   ____ e) None of the above

8. When operating near an open dock door, what is the minimum safe distance to remain away from the dock door?

   ____ a) 2 to 3 feet

   ____ b) 10 feet

   ____ c) 3 to 4 feet

   X d) 6 feet

9. When crossing railroad tracks, a forklift should be driven across the tracks in what direction?

   ____ a) In reverse

   X b) Diagonally

   ____ c) Forward, but slowly

   ____ d) Operator's choice

10. Persons may be raised in an approved work basket if?

____ a) The operator remains at the controls at all times

____ b) The basket is secured to the forklift

____ c) The person is a member of the maintenance department

____ d) If the forklift is inside the building only

__X__ e) Both a and b

____ f) Both c and d

True or False Questions - Four points each

1. __X__ True ____ False - Riders are never allowed on a forklift.

2. ____ True __X__ False - All forklifts has the same lifting capacity.

3. ____ True __X__ False - The OSHA speed limit for forklifts is 5 m.p.h.

4. ____ True __X__ False - Inclines do not affect the center of gravity of a forklift.

5. ____ True __X__ False - Pedestrians must yield to forklifts.

6. __X__ True ____ False - Operators must always wear their seatbelts.

7. __X__ True ____ False - A forklift cannot be parked closer than 8 feet from the center of a railroad track.

8. __X__ True ____ False - The load capacity of a forklift is based on a 24 inch load center.

9. ____ True __X__ False - Centrifugal force has no effect of the forklift's center of gravity.

10. ____ True __X__ False - Forklifts weight enough that wet surface conditions do not affect them.

Fill In The Blank Questions - Four points each (one point per blank)

Note: The exact wording or terminology is not the critical element in this testing. If the answer given properly conveys the information it should be considered correct.

1. If a forklift is tipping over, the operator should not jump. In addition, to further protect themselves the operator should;

   a) Hold tight to the steering wheel

   b) Brace your feet

   c) Lean away from the impact

   d) Lean Forward

2. When parking a forklift the operator must be certain what steps are taken?

   a) Forks lowered to the floor

   b) Engine In Neutral

   c) Key in the 'off' position

   d) Parking brake set

3. When operating on an incline the forklift operators should reduce SPEED and make no TURNS to prevent the center of gravity from shifting. The forklift should be DRIVEN up the incline, and BACKED down the incline.

4. When loading or unloading a truck/trailer the forklift operator must complete the following preliminary procedures. They must insure the wheels of the trailer are CHOCKED and place a NOSE JACK under the front of the trailer if no truck is attached. Inspect the trailer FLOOR to be certain it will support the weight of the forklift and the

load being carried. Be certain the dock plate is **SECURED** before driving over it.

5. Forklifts can be powered by a number of different types of engines/motors. These include;

a) LP Gas

b) Diesel

c) Gasoline

d) Electric

Chapter 15

Review Process

The trainer will review all sections of the hands-on training with the trainee. All questions and concerns will be addressed in detail. Any areas which the trainee feels have not been adequately covered, or does not feel comfortable performing will be reviewed and additional practice time arranged.

Once the trainer feels the trainee is capable of operating the forklift in a safe manner the trainee will be allowed to perform regular tasks within the work environment. During this time the trainer will continue to observe the trainee and immediately correct any errors and address questions and/or concerns.

The observation and training period will continue for a maximum of thirty (30) days. When the training period ends, the trainer will complete a hands-on evaluation of the trainee's abilities in the workplace. This evaluation will determine if the operator is to be certified as a forklift operator.

Chapter 16

Forklift Operator - Hands-On Evaluation Check List

Checklist is to be completed by a competent person familiar with the safe operation methods and procedures of forklift operation.

Evaluation Completed by: _____

Operator Being Evaluated: _____

Date of Evaluation: _____

Areas evaluated

Properly completes daily inspection of machine and completes daily check list: _____

Demonstrates a firm understanding of the various components of the forklift: _____

Properly adjusts equipment before operating forklift (seat, mirrors, etc): _____

Wears seatbelt and any other required Personal Protective Equipment: _____

Operates machine at a safe speed : _____

Makes turns smoothly and at reduced speeds: _____

Sounds horn at intersections and when approaching pedestrians: _____

Keeps load as low as possible, normally 3 -4 inches off the ground: _____

Knows safe and proper refueling method for the forklift they are operating: _____

Is familiar with basic controls of the forklift: _____

Follows proper procedure when parking the forklift: _____

Watches for overhead obstructions: _____

Backs down and drives up when on an incline or ramp:_____

Follows proper safety procedures when preparing to unload trailer:_____

Operates the forklift slowly and carefully when loading/unloading trailer:_____

Has an overall good understanding of how to safely operate a forklift:_____

Chapter 17

Basic Daily Checklist

Check List Completed By: _____

Date Completed: _____ Forklift Number : _____

Checks to be completed before starting engine;

____ Inspect tires for excessive wear or separation.

____ Check hydraulic hoses for leaks, cracks, or wear.

____ Sound horn.

____ Check all controls, be certain they return to neutral position.

____ Inspect forks and fork carriage for damage. Be certain forks are locked in position.

____ Complete walk around of machine for obvious damage or defects.

____ Check seatbelt for proper motion and excessive wear.

Checks to be completed with the engine running;

____ Check parking brake.

____ Check all lights on machine.

____ Raise the mast to check for smooth operation.

____ Tilt and shift the mast to check for smooth operation.

_____ Check brakes for proper operation.

_____ Check steering for excessive play

_____ Pull forward and check for leaks where forklift was parked.

Other item which needs attention: _____

Notes:

_____

_____

_____

_____

_____

_____

Chapter 18

Extensive Daily Checklist

Check List Completed By: _____

Date Completed: _____ Forklift Number : _____

Checks to be completed before starting engine;

____ Inspect tires for excessive wear or separation.

____ Check hydraulic hoses for leaks, cracks, or wear.

____ Sound horn.

____ Check all controls, be certain they return to neutral position.

____ Inspect forks and fork carriage for damage. Be certain forks are locked in position.

____ Complete walk around of machine for obvious damage or defects.

____ Check seatbelt for proper motion and excessive wear.

____ Check engine oil level.

____ Check coolant level.

____ Check hydraulic fluid level.

____ Check brake fluid level.

Checks to be completed with the engine running;

____ Check parking brake.

____ Check all lights on machine.

____ Raise the mast to check for smooth operation.

____ Tilt and shift the mast to check for smooth operation.

____ Check brakes for proper operation.

____ Check steering for excessive play

____ Pull forward and check for leaks where forklift was parked.

Other item which needs attention: _____

Notes:

_____

_____

_____

_____

_____

The information provided in this manual is of a general nature, based on certain assumptions. The content may omit certain details and should not be regarded as advice that would be applicable to all businesses. If seeking resolution of specific safety, legal or business issues you should consult your safety consultant, attorney, business advisors, or your local OSHA office. The information presented is not a substitute for a thorough survey of your business or an analysis of the legality of your business practices. The information provided in this manual should not be considered legal advice.

Printed in Great Britain
by Amazon.co.uk, Ltd.,
Marston Gate.